WHEN THE SHOOTING STARTS

What to Do When a Gunman or Terrorist Strikes

Book design by:
Arbor Services, Inc.
www.arborservices.co/

Printed in the United States of America

When the Shooting Starts
Travis Bruyer

1. Title 2. Author 3. Self-Help

Library of Congress Control Number: 2021906779
ISBN 13: 978-0-578-88594-0

WHEN THE SHOOTING STARTS

What to Do When a Gunman or Terrorist Strikes

TRAVIS BRUYER

Contents

A Word from the Author

Motivation is the key to unlocking the significance behind the infamous why? If motives fail to be explored, pursued, and discovered, it will be challenging to reach the purpose. Without this important piece, the puzzle remains an enigma—unsolved and unresolved. Awareness of a person's why—his or her reasons for a specific action—leads to an understanding of the result, why a person did what he or she did.

Everyone has a backstory wrapped within a why shell, a story begging to be told. Why? Because it is an individual's testimony, complete with experiences others may learn from. Every footprint is part of a legacy. To have value, every legacy must be brought into the light.

Although my story reads as different, in reality it isn't. My experiences throughout the years have both frightened and saved me—saved me because I had the knowledge to identify, pinpoint, and manage situations, potentially dangerous events that could have easily turned tragic. Consequently, I felt inspired to write *When the Shooting Starts*, a commentary that targets everyone with a purpose and inclination to act in the best interests of our children and community.

Dedication:

To all who believe as I do that life is a most precious gift, to be cherished, preserved, defended, and protected!

Introduction

A book is deemed a worthwhile read and a crucial source of information when the author's credentials are considered valuable based on experiences judged a font of empowerment.

A renowned liaison and law enforcement instructor, Travis Bruyer is known from coast to coast for his commitment and die-hard collaboration with community involvement. During his quarter-century active-duty career, he exercised a major role in the successful outcomes of countless categories of criminal investigations including but not limited to search and rescue, immigration, drug interdiction, vessel recovery, and boating inspections. Mixed in were a broad spectrum of violent and nonviolent unlawful acts.

As a law enforcement officer, he continued that success when he was hired and assigned to patrol, and later worked as a tactical instructor, detective, and special assignments, in addition to hostage/crisis negotiations and coroner duties. Bruyer's expertise qualified him to instruct on the appropriate handling and utilization of tasers, batons, pepper spray, defensive tactics, and the employment of force in managing both domestic violence cases and armed situations.

Currently, Travis is the executive director of the Montana

Institute for the Arts, a 501 © (3) nonprofit film and art organization that introduces and integrates students into film and television courses. A script consultant for film projects, he provides military and law enforcement training for actors and writers. Mr. Bruyer also shares his expertise as a licensed private investigator, crime scene consultant, and teacher of public safety education.

Acknowledgments

My thanks and appreciation to "You!"

Yes, each one of you who will read my words.

Why?

Because you have wisely chosen to step away from, to decisively refute the victim role.

Because you have committed to the preparation of a family member, friend, or random citizens for either arbitrary or deliberate acts of violence.

My intention for writing *When the Shooting Starts* is not to terrorize or frighten you. I do understand how my words may be misinterpreted—but I ask you to read my book a second time, and you will understand my wish for open discussions on safety and preparedness for sudden unexpected danger. Knowledge reassures; ignorance frightens. Having appropriate insights and smart judgment together with familiarity and "know-how" lead to safety. Being unprepared, unqualified, caught off guard, confused, and ill-equipped lead to tragedy. My purpose is always safety. It guides my career choices and propagates my reason to be.

Thank you to my friends and family, my wife and children

for the love, inspiration, and empowerment you impart every day, as well as my mom and brother for their unwavering encouragement. Without your loyalty, patience, tolerance, and a shared belief in the positive impact of what I do and what I write, this book would not have come to be.

CHAPTER ONE

Part I
Why I Wore the Uniform

"I came, I saw, I conquered."
~ Julius Caesar

The 2010 movie *True Grit* opens with a poignant quote from Proverbs 28:1: "The wicked flee when no man pursueth." I have to admit, the film started with a brisk, curt shakedown, not only thought-provoking but action stimulating. Repeating the biblical citation in my head brought me back to a time when my involvement in law enforcement was at its peak. Ultimately, it stirred memories of a life-altering event in my life, a memory that never really slipped behind the lowered curtain along with my relentless search for answers.

"The wicked flee!" Why? Why run? Why hide? In the end, not even Houdini, the world-renowned illusionist and escape artist, was able to flee his failings. There is no successful flight from the reflection in the mirror, the conscience—truth.

Who has been spared the role of villain in someone's story? Certainly, very few can honestly boast, *I have.* To those unable to shout, *I have*, I pose the question: What was your reaction as a nemesis? Was it confrontation or avoidance in the face of wrongdoing? Did you turn a misdeed more treacherous via denial, or did you attempt remediation through amends by admitting fault? Evocative thoughts to ponder.

August 22, 1989

A last-minute decision to satisfy one of my father's bald-faced options for my future, centered on either a college education or a military experience, motivated me to join the freshman class at Montana State University to pursue studies in film and TV production. The criteria for the academic alternative included an eventual departure from the family nest, sharing living arrangements with Hank, one of my best high school friends, and my penchant for producing and directing backyard movies.

The hearsay circulating freshman life, namely all-you-can-eat cafeteria offerings; the ease for successful, albeit illicit procurement of beer; and the large gatherings for sporting events uniting the entire community turned out to be a much more appealing life path than the military.

My transgressions, however, were not limited to underage drinking but widened to comprise the unlawful inclusion of a toaster in the dorm. Why sin for a toaster? In a word, Pop-Tarts—to satisfy my cravings for the mouthwatering confection with warmed sugary fillings ranging from peach, blueberry, and lemon crème to chocolate, cinnamon, and the frosted variety lineup.

Since rarely can anyone escape the truth, I was caught. A solution-oriented man, I negotiated with the dorm residential advisor, bartering the toaster surrender for his intervention in convincing the metalheads next door to tune down their ear-piercing REO Speedwagon morning wake-up call. My bargaining chip was precise—no alarm silencer, no toaster surrender. Blasts of how someone heard it from a friend and someone else heard it from a friend … continue to haunt me. I moved on.

The college radio station posted an opening for a DJ. Believing I was qualified for scratching vinyl like a professional, the opportunity piqued my curiosity. Simultaneously, the job description specified a hip-hop repertoire. In possession of every "G-rated" rap album since '84, I met the prerequisites and was hired.

Between the hours of nine p.m. and one a.m., my new identity was "DJ Travitron," spinner of the dubbed G-rated classics featuring the 80's toughest gangsters and rappers: A shout-out to my favorite, "Big Daddy Kane," aka Antonio Hardy with his Good Nda Hood, "Peace Out."

After two p.m., Hank took over to spin classical music. Quite an impressive expert, he was familiar with everyone and tutored me for my music appreciation final. I studied Frederic Chopin, Wolfgang Amadeus Mozart, and George Handel, to name a few, although I will never forget Giaochino Rossini's "Il Barbiere di Siviglia." Oddly, the comic opera, or *opera buffa* as the Italians say, triggered memories of the "Rabbit of Seville" starring Bugs Bunny and Elmer Fudd: So much for cultural appreciation and the acquisition of refinement.

May 14, 1990

Tuesday, May 14, was similar to May 13 and the sequential previous twelve days—somewhat unseasonably chilly though uneventful. In all probability, May 15 would follow in its footprints. No, ignorance is far from bliss. At age nineteen I was oblivious to my surroundings and walked from class to class more often than not in robotic mode. My Photography Lighting 101 class had concluded, and I headed back to my room for a brief nap before DJ Travitron surfaced.

May 15, 1990, 1:00 a.m.

Several minutes before the end of my shift, Hank arrived to spin the classics. "Start spinning," I teased. "I'll see you back in the dorm." How could I have even minimally imagined what would happen next?

Part II
The Shooting

May 15, 1990, 2:30 a.m.

I answered on the first ring. Hank's voice sounded agitated. "Are you okay?" Just moments prior, a loud sound had reverberated in the early morning stillness, fully awakening me.

Now gazing out the window, I noticed several cop cars in the parking area, lights flashing. "Yes. What's going on? The cops are all over the campus."

"I received a teletype from the associated press at the radio station. I promised to call him back."

Walking to the door, I pulled it open vigilantly. Whatever follows a question mark was to be threaded cautiously. Add men in uniform, and apprehension brews. I poked my head out the door without removing my hand from the handle to allow a quick closing if need be. Several other students shared my curiosity.

The corridor was clear, peaceful. Suddenly an odor of lingering gunpowder penetrated my nostrils. *This can't be good*, I thought. A few guys stepped out and began heading for the stairwell. Trailing behind, I noticed a police officer had just finished draping yellow caution tape across a doorway.

Continuing down another level, I paused in the entryway to

the first floor. A continuous trail of blood from a dorm room to the lobby caught my attention. Several slow, deep breaths were a successful prelude to centering my thoughts. *What happened?* I wondered. The setting looked dreary. Unable to speak, I gazed at another student standing in a doorway.

The resident advisor had discovered a student who was the victim of a shooting. Stunned, I blurted, "Shot!" *What the hell is going on?* I questioned in silence. *How can this happen?*

Identified as James, the nineteen-year-old lay gasping for air in the hallway while the advisor waited impatiently for the arrival of ambulance and paramedics. The pungent aroma of gunpowder seemed like a concoction of sulfur mixed with the chards from overly seared meat. In a rush to prevent a panic, campus maintenance and police were at work cutting the carpet to remove traces of the blood-soaked trail.

Eventually, the victim was fully identified as James Clevenger, a teenager from Billings. Several hours thereafter, he took his final breath at Deaconess Hospital in Bozeman. Time of death: 5:15 a.m.

Police were swarming the campus. The shooter had not been apprehended, thus the possibility of a campus presence with an encore shooting remained a frightful reality. Consequently, lingering in the hallway was not a wise option. Word spread quickly: Another nineteen-year-old student, Brian Boeder, had been shot in his room.

Transferring to Montana State from the American Free Lutheran

Bible School, my primary objective had been to complete theology studies to undertake a career as a youth pastor. I had previously spoken to Brian, a young man from Plymouth, Minnesota.

I visualized his room, number 130, sealed off with caution tape like James's. Shot in the arms and legs, Brian's major artery was likely pierced by the bullets, ending his life. Time of death: 4:45 a.m.

Police reports post-investigation based on the eyewitness testimony of a jogger stated that a campus officer, Rex Duncan, claimed to know the identity of the shooter. At the exact moment of the crime, the student jogging past the scene shouted, "Call the police! There's been a murder!"

The words of the jogger were, "There's been a murder." Wouldn't the obvious cry have been, "There was a shooting" or "Someone was shot?" But he specified "murder."

Let's examine the term murder. *MURDER (noun), pronounced mur-der: "To kill a human being unlawfully and with premeditated malice."*

May 15, 1990, 4:50 a.m.

Sheriff O'Reilly of Lewis and Clark County sent out a bulletin: One of his deputies, who had engaged in a high-speed pursuit, identified the vehicle from the BOLO (be on the lookout) broadcast issued from the law enforcement agency. The report was clear and concise:

the vehicle had crashed through a fence, a gas pump, and a power pole. The action appeared to be deliberate.

Thankfully, the culprit was apprehended. The campus student body and professors could breathe easily, or could they? In reality, it was the epilogue of the beginning, the beginning of tears and grieving, the beginning of the unanswerable question why, and the beginning of the remembrance of the young men who had met their forced destines, orchestrated by criminal design.

Remember, with most violent crimes there is usually preparation. In this case, an investigation determined that the assassin had stalked the corridors of Langford Hall, consumed an excessive amount of liquor, and previously damaged a parked vehicle after driving recklessly in his pickup. If he had any inkling of conscience, he may have realized his life would be severely compromised when his father learned of the road rage incident. It was just a matter of time. One certainty remained—like the pendulum of a clock, he likely feared he would be directed toward enlisting in the military.

All it took was a determined, self-assured knock on the dorm room door.

"Come in," Brian called out, inviting the visitor to enter. A slight, barely audible creak announced the countdown of seconds before—the prelude to four piercing shots from a sawed-off Marlin shotgun. Lunging to intercept the intruder, Brian was wounded, knocking the gun to the floor. The suspect fled, leaving Brian questioning as death defeated life. *Who would want to do this to me?* There was no answer.

May 20, 1990

I planned to visit Carroll College to spend the weekend with a friend of mine. Unforeseen circumstances resulted in a sudden cancellation. On that Saturday, an armed man walked into the cafeteria, opened fire, and ended the life of an employee while seriously wounding another. The report stated: random attack. I phoned my parents and informed them I was coming home.

I enlisted in the US Coast Guard.

March 25, 2003

Thirteen years later I was in law enforcement, a deputy. Though many view the town of their growing-up years as a setting for childhood memories, boring and stilted without any promising future career options, I found mine a place conducive to raising a family. I also realized the wise foresight my parents exercised in settling there.

The event at MSU would have remained a traumatic experience for any nineteen-year-old. Indestructible in my mind are the visual and auditory sensations. The odor of death invading Tuesday, May 15, 1990, permeates, never fading, not even on the windiest winter days. It persists—impacting, affecting, and disturbing, regardless of the passage of time. Powerful, like a time machine, it transports me back to my freshman year at MSU.

Terrified of the unknown, afraid of teenagers' lives being annihilated in a few seconds, and fearful of living my day-to-day life eventually led me to accept that it was okay and understandable to be frightened. All unknowns are enemies of tranquility. Why do people sabotage and take the lives of others—children, teens, men, and women? Why? And more importantly, I never again wanted to be a passive observer.

In 2018 I introduced myself to James and Brian's still grieving families, mentioning I too had been a freshman at MSU when the tragedy occurred. Both families continue to petition the Deer Lodge Prison Parole Board when the shooter is up for his hearing, to contest a release from prison to parole supervision to serve out the remaining sentence.

These families have no happy Mothers or Fathers Days, no celebratory birthdays, and no festive holidays with their departed children. Instead, they have painful, frustrating questions and suppositions focused on what their sons would and could have accomplished in life, if only … This is why I wore the uniform!

CHAPTER TWO

To Confront or to Run—That Is the Question

"Do what you can with all you have, wherever you are."
~ Theodore Roosevelt

"I am sending you out like sheep among the wolves. Therefore, be as shrewd as snakes, and as innocent as doves."
~ Matthew 10:16

More often than not, scriptural passages can drive home meaningful messages without sermonizing shouts from street corner preachers. All it takes is a few minutes to browse a passage to uncover an important tip for a better, more secure life.

The encouragement is to be "as *shrewd* as snakes." This in turn translates to being observant, vigilant, and aware while present in the moment. What exactly does this call to mind? The most obvious response is situational awareness, a strategy I can never overemphasize.

Perhaps the key here is sustained alertness and attention to detail.

Although this Bible passage was written almost two thousand years ago, it describes today's world, a world that may present a life-altering or worse outcome, a hazardous place for kids and/or adults.

With today's schools sometimes tragically turned into battle-grounds, children should be cautioned about the potential evils that lurk around friendly classrooms and cafeterias. This is why I favor knowledge over scare tactics, which I believe result only in panic reactions and emotional/physical momentary paralysis. Scaring children implants a terror button in their minds. Terror can be a prelude to a fatality, as it erases all focus and concentration on the situation at hand.

So, what is the best way to educate kids wisely? Indulge in a be-with moment before leaving your home. Gather your wits, your intelligence, your common sense, your intuition, and the survivor skills acquired from those in the know—law enforcement professionals with training, experience, and a nose for danger. Everyone, when they leave the sanctuary of their homes, wants to return without indelible battle scars. They want to return unscathed, both physically and psychologically. Therefore, the lessons taught must be age appropriate yet precise.

Be sure to include a discussion of basic risk-assessment skills. Sometimes a risk nets a benefit; other times, a tragedy. For this reason, I support training from an early age. Learn not only the ABCs but the SASs—situational awareness strategies!

I underscore the importance of never making rash judgments and verifying before giving a place or person a clear status. Why? Because appearances can be deceiving. An individual wearing a long, heavy, dark jacket may or may not represent a red flag. A young child who crosses the path of such a person may feel anguish, whereas an adult knows how to verify an oncoming peril. A long, heavy, dark jacket in December is trendy whereas a heavy jacket in late July could represent danger.

Instruct kids to employ all their faculties—vision, hearing, sense of smell and taste, and ability to feel. Often these are the most reliable criteria for good decision-making. I tell everyone that awareness is simple—if it looks, smells, tastes, feels, and sounds like danger searing on the fire, trust your senses and avoid circumstantial distractions. Instead, be watchful and ready to take the appropriate action. Don't be disabled; be enabled, and advise every child to adopt this motto.

I like the phrase "trust until it becomes impossible to continue, then proceed with caution." Be sensible, relatable, and practical—and question if necessary. Protection, however, cannot be guaranteed from a distance. After the Columbine mass shooting, antennae rose: What if this happens again? How can such a devastating loss of lives be thwarted if not completely prevented?

Greg Crane, a Texas law enforcement and SWAT officer, in collaboration with his elementary school principal wife, formed ALICE (Alert, Lockdown, Inform, Counter, and Evacuate). Initially labeled "crazy" for their suggestion to train young schoolchildren to overcome an armed intruder, the provocative and contentious officer and principal pursued. Knowledge is power. A colloquial catchphrase? Perhaps. But on the flip side, a proven winner.

I am not endorsing this program as the best for your school, but my experience has taught me that what saves lives is a plan focused on knowledge of tactical awareness in conjunction with practical information on quick, efficient physical defense responses.

However, a silent issue remains—attendees often don't take the training seriously. Why? Because in reality people attend, convincing themselves their presence satisfies a requirement. And business/agencies can fulfill their obligation to document they provided employee training. Sadly, people often seek training simply to check a box.

Every business, organization, and institute should pose the question: do we aim for commercial success, or do we wish to provide effective training? Accomplishing both can be challenging.

Situational awareness training is comparable to the scene at an automotive dealership. If a customer enters and requests a sedan, the salesman will list the safety features of a truck as a better all-around vehicle. Nevertheless, the customer insists on the sedan. If the salesman wants to make a sale, he will sell the customer the sedan he wants. The decision is based not on what is best and safest to drive but on the customer's want. All information was cast to the wind. In a sense, a similar strategy has misrepresented and diluted martial arts training.

Yet the situation worsens. Statistics chart over 130 fatal shootings in the United States since Columbine—twenty-eight alone in the Sandy Hook 2012 shooting. Alarming? Denial cannot negate the truth.

Therefore, though ridiculed, Officer Crane and his principal wife introduced the armed intruder nemesis into the curriculum to teach the children how to survive or at minimum to prolong survival until help could arrive. Above all, they sought to ward off panic and its deadly consequences.

With the *imprimatur* of the FBI, the Run Hide Flight program presented by the City of Houston and Homeland Security somersaulted the shelter-in-place regimental rules and guidelines and opted instead for a more action-oriented reaction. This program encouraged taking flight then acting aggressively should a potential victim find themselves in a confrontational face-to-face with the armed intruder.

Alerting children to the possibility of armed intruders will certainly cultivate an awareness; however, be advised that it can also produce undue stress and apprehension if not presented wisely.

Perhaps the visual approach is best, presenting the information through videos that demonstrate easily applicable options for survival, strategies that children can learn and master. Encouraging practice drills solidifies understanding of the material and increases self-confidence. The familiar always removes the anxiety of the unknown. Knowing what to expect and how to handle it should be the goal.

Years ago, school drills were scheduled for fires and even tornadoes in certain prone areas. Today we have a lockdown drill, and every child, including the youngest, should not only be aware of the difference but be able to answer the question: What happens when word comes that an intruder has entered the school? The first step involves an adult identifying the intruder and establishing whether it is one person or multiple.

Once the danger has been pinpointed, the options are: lockout; lockdown; staff lockdown intervention; shelter in place; hold in place; evacuate; and run, hide, & fight. I don't espouse a hard, fast line for any one of these options, as I believe reactions should be different depending on the age of the children and the nature of the situation. The best option is to remain flexible. Once the intruder alarm is given, the staff/administrators will determine the course of action.

After a series of dire shootings, most schools have adopted a similar protocol. Before an event occurs at your children's school, it is wise to inquire who has received appropriate training. Parents entrust their children not only to qualified instructors but knowledgeable administrators capable of protecting the vulnerable in the event of an emergency. Sadly, school violence is not new and is no longer a rare occurrence.

"Evil is powerless if the good are unafraid."
~ President Ronald Reagan

Fear has the power to destroy, just as powerfully as an armed intruder in front of an ignorant, panic-stricken target. Yet knowledge detonates fear. Let's circumvent this preventable scenario by examining some strategies once the location of the intruder is determined.

Lockout: In the event of a suspicious person or inkling of an apprehensive, questionable, out-of-the-ordinary situation, an order is given to secure the building from outdoor intruders. Students, teachers, and staff remain in the building and proceed with daily activities. First, teachers lock all doors and windows and take roll call for a headcount of all who are present. Then they continue the planned lessons, aware that a lockout is in effect.

Lockdown: When a previously suspicious individual becomes a viable threat on the building grounds, students are ushered into classrooms and secured behind locked doors. They are urged to seek refuge away from windows or doors, which are targets for shooters. Lights are switched off and all phones turned to vibrate mode and left dormant. No communication is allowed until law enforcement arrives.

The intruder, hunting for prey to kill, may have already gained access and roams the premises. It is safe and wise to assume the intruder knows that law enforcement has been alerted. Remember, they are on a shoot-to-kill mission with neither respect nor concern for their own safety. The goal is to take victims by surprise and open fire. Consequently, locked doors may be deterrents.

Staff lockdown intervention: Here is where training pays off. Competent knowledgeable staff quickly disengage doors, shoving bookcases in front to impede smashing them in. When available and determined to be safe, ceilings and large pieces of furniture can be used to conceal children. Or children can escape through windows if police personnel are visible. These individuals often risk their own lives to save their children.

Shelter in place: This option is adopted to protect from natural dangers such as tornadoes and other disaster-provoking weather. Students are called to silence and advised to remain in their classrooms or to retreat to a preestablished designated hazardous-weather refuge. This approach is adapted to limit exposure to the elements. Teachers will attempt to assure student safety.

Hold in place: If the occurring situation is defined as minimally perilous or questionable, all foot traffic in the building is curtailed.

Anyone in a corridor, restroom, or hallway should immediately return to the classroom.

Evacuation: If the scenario involves an armed intruder in the building or a rapidly spreading fire, students, teachers, and staff should not panic (this is why I suggested practice drills) and proceed to the exits in a quiet, methodical manner without dillydallying or faltering, while staff monitor to be certain no one is left behind. The focus should be on saving lives, not possessions. Staff should leave disorder behind but slip their cell phone in a pocket. I emphasize: do not go into fire-drill mode. The intruder may pull the alarm, anticipating a group target to fill with bullets.

Run, hide, & fight: If hiding is not an option and other alternatives are nonexistent, leaving the premises as soon as possible is best. Should fighting be your chance for survival, teach children to do so without honoring any code of ethics. Yes, they should fight dirty if need be. Lives are on the line. It's a toss-up between you and an intruder, so use the swarm (gang up on them) method. Use this option when you have no other chance and the proximity rule is in effect.

Let's assume you open the door and surprise an intruder, gun in hand, finger on the trigger or loading for a bloody foray. You have two options: 1) retreat, lock the door, and call 911, or 2) kick the gunman right between the legs, then once he is doubled over, poke him firmly in the eyes with the intent to leave him staggering in pain. YOU steal the fight. Hit, slam, bang, and stick the gunman with whatever is available—fire extinguisher, broom handle, book, pointer, or laser beam. It will catch the intruder off guard. Why? Resistance is not in a shooter's mindset. With a gun in hand, a criminal feels powerful and undefeatable, as if immune to any violence, defense, or reprisal.

This is why I endlessly repeat the importance of educating children concerning their options. A consequence of deflection of duty in this camp could put them in a situation where they are directed by a staff member who (fearing for their job) could make an erroneous policy-based decision instead of one designed to save lives. Yes, experience has taught me such decisions have occurred at the expense of many victims, a majority of them innocent children.

CHAPTER THREE

School Policymakers & Administrators

"Science and policymaking thrive on challenge and questioning; they are vital to the health of inquiry and democracy."
~ Nicholas Stern

Prioritize Safety

As a rule of thumb, I emphasize religiously how preparedness and elevating awareness are the cornerstones of many safety policies. Carrying it a step further, I'll say that one of the most successful ways to deal with varying categories of intruders, be they armed or unarmed, is by way of training and the implementation of knowledge. Teachers and staff should be trained via exercises to handle emergencies, because they hold the safety of the children in their hands.

The options are: Be safe or be regretful. More often than not, ignorance and carelessness kill just as quickly as gunfire or knife stabs, and situational awareness saves lives with the same reliability as medication for cardiac issues.

Remember, what we either do or fail to do has consequences, and consequences last forever. Never forget that a survivor or a victim is created every ticking second. The 911 call should be made immediately in the face of any threat, as the police response generally takes about ten minutes. Meanwhile, students and teachers should not be in fear-panic mode. Never scare children; such a strategy could result in what I define as a trapped-in-an-ice-cube mode in which students remain frozen and powerless. On the other hand, a rational dose of fear warns of suspicious threats and possible danger. The correct balance of fear and seriousness is necessary.

We must accept the fact that there is no expenditure of energy without the possibility of blunder and inadequacy. However, fear of making a mistake should never be a deterrent to intelligent action. The truth is dependent on knowledge and information. Replacing exaggerated ignorance-based fear with training and information, practical common sense, and the ability and courage to decide what to do may allow potential victims to skirt a tragedy.

Fear should not remain after effective training has taken place, clearing the way for safety. Consequently, the onus on educating and thus protecting students rests on school administrators, the policymakers responsible for curriculum content.

Shooter Revelations

In the words of Marie Curie: "Nothing in life is to be feared. It is only to be understood." How true: Just think back to mass school shootings. Is it possible to deter an attack? Law enforcement reports show that decisive, action-oriented intruders often demonstrate predictable behavior patterns weeks before the actual attack. Attention to detail, together with situational awareness on the part of school administrators and policymakers, can spot this behavior that can signal imminent violence. Thus criminal activity, which unfortunately is seldom documented for the appropriate authorities to take action, can be identified.

Here is where training and an experienced eye come into place. In many school shootings, the armed intruder did little to conceal his intentions before executing his plan. However, angry, aggressive, seemingly violent behavior on the part of a student does not always escalate to become a mass shooting.

School administrators and teachers are dealing with adolescents in transition from children to adults. Frequently, students' bodies are in chaos, and their willingness to unburden their minds wavers due to fear of authority or judgment. As a result, pent-up, tightly capped unresolved conflicts can implode. One isolated loss of control cannot in all sincerity determine that the teen is a potential mass shooter on the road to a bloodbath. Nevertheless, frequent repetitions should send up a red flag. Administrators should be aware of questionable or worrisome behavior and document it.

Let's assume you're seated in the jury box. The prosecutor

introduces the defendant's charge—grand larceny for having stolen a wallet. No mention is made of the intent to steal. Video footage is offered to seal the charge—footage that reveals an attempt to return the wallet at first, then a decision to keep it as a means of survival.

The perpetrator has no prior history of stealing and attempts to rationalize his behavior, saying he lost his job due to layoffs and is thus not in a position to meet his power and light payment. An opportunity presented itself and he seized the moment without premeditation.

Let's change the protagonist in the same courtroom setting. This time the prosecution introduces the wallet thief, citing priors for auto theft, drug possession, the default of child support, arson, and animal cruelty. Would you judge the two men equally for the same crime? Of course not. One exhibits a circumstantial tort and the other habitual.

Emphasizing the importance of documenting behavior is the *modus operandi* of law enforcement. Reports are written, citations issued, and a paper trail paves the road ahead.

Moving to the campus, due process of this same nature is valid for students who engage in disruptive violent behavior, who frequently blame others, and who have a history of bullying both on and off the school grounds. Further revealing tells involve gallows humor, depression, and being a victim of abuse. This type of individual rejects accountability and refuses to accept responsibility.

Here we have the identikit of a troublesome individual to be kept under observation.

Educators—Where Is the Starting Point?

The academic year is not only about the annual syllabus outlining the course work at various levels but about providing a safe educative environment for optimum growth and learning. Troublesome students waving invisible danger banners should have the opportunity to redirect their negative energy onto worthy endeavors. Much as I dislike admitting it, it does take a village.

How can school administrators and policymakers alter the tide? Assuring students' and teachers' safety at all times requires planning and commitment. As a start, familiarize all school personnel, both administrative and students, with the process of evacuation and lockdown. Though far from an easy task, furnishing a productive nonviolent educative climate with assurances of safety is possible. Attaining this goal requires open lines of communication involving the full academic staff, including counselors, coaches, teachers, administrators, and maintenance.

Include parents, students, and of course law enforcement in the jurisdiction. Be certain to solicit the expertise of law enforcement personnel for training and quick response. Prepared with a wealth of knowledge, their advice and resources often curb tragedies in the bud, thus saving lives.

Eventually, fire/medical personnel should be included in instructing students about surviving an armed intruder attack. However, for the moment I will postpone discussing their roles. Why? Because the fire department tends to adhere to an open-door policy based on the concept to let it breathe—feel the fire and determine the heat source. Characteristically they throw open windows, push open doors, and park in front of everything, in direct contrast to law enforcement's more discreet closed policy.

Although I speak in the first person, my views reflect general police strategies. Our vernacular encompasses terms like "tight perimeter," "eyes on all sides of the building," and "blocking exit routes." Our policies differ. Police vehicles are tactfully parked away from the facility to assure cover and concealment. Though both law enforcement and firefighters are focused on one goal, the approaches vary. Safety, however, is the unconditional priority.

Eventually, a training day or tabletop exercise may be best, which would include structured meetings during which personnel gather in a study or classroom to examine and explore specific positions and responses to a precise emergency, an exercise similar to role-playing. Firefighters, LEO (law enforcement officers), hospital staff, mental health professionals, local government, and the media are typically included in these sessions.

Incident Command Systems (ICS) will be involved in emergency preplanning. This is a homogeneous tactic dealing with the command, control, and coordination of emergency first responders, making available an operational pyramid that allows responders from numerous agencies to be effectively operative.

Once all the essential players are in place, the importance of the exercise will be recognizable as well as the indispensable collaboration of each participant. Educators will be on the front lines managing media perceptions, reporting, and catching clues from law enforcement and medical personnel to report headline news as it breaks.

Should educators, administrators, and policymakers require guidance due to indecisiveness regarding the best approach, there are over eight thousand nonprofit antiviolent school programs available to choose from that offer instruction to law enforcement, who in turn train school staff. A collective endeavor, it is a community-organized commitment to the safety of children.

Security cameras and/or assigned police as school resource officers form an excellent grouping. Unfortunately, finances always enter the discussion, as money is pitted against the safety of children. Thankfully there is a solution: a contract between the police department and the school is drafted to offset funding and ensure an officer can be present in the school. A noteworthy program, this alliance permits the officer to familiarize him/herself with staff and youth and form a partnership for collective protection.

Creating Barriers

Access control is one of the most pressing issues centered around campus safety. Locking doors between the hours of 8:15 a.m. and 3:15 p.m. is not very practical unless educators and law enforcement can secure parental agreement. Most moms and dads shun the idea of restricted entry to visit with kids at will. This mindset has to be readjusted. Better a controlled entry policy than a harmed child.

Parents must realize the gravity of the situation and the potential calamities arising from disorganization, ignorance, and a slipshod school protocol. Schools that lock doors from random intruders provide the first line of defense. After 8:15 a.m. a visitor should be buzzed in once a staff member has reviewed the camera. From the official entry foyer, the visitor should proceed to another layer of doors and windows (preferably retrofitted with a sturdier, shatter-proof glass). At this juncture, a staff member would question the individual to determine their motives for the school visit. From here barriers can be increased or decreased depending on the subject's request. By announcing the visitor's name or summoning other staff members, a chain reaction is set in place, initiating a cautionary order to clear hallways and lock classroom doors.

Facilities with open campuses modeled after colleges may find this protocol challenging when students have the freedom to roam at will. Most campuses are not settled behind ivied walls, thus often random pedestrians stroll aimlessly. Consequently, administrators and educators may easily lose sight of who is on the property and why.

The issue of keys always surfaces—far too tempting for a disgruntled, mal-intentioned employee to make copies. Individuals with good intentions may misplace or lose keys, rendering the situation problematic and difficult to monitor. How can they be tracked? They can't unless there is a deterrent barrier such as visitors' badges and a key sign-out procedure.

In the event of key/card access to communal areas, the school is assured of an auditing/tracking system with a date/time stamp displaying the identities, not only of who is and was on the grounds but the exact location of the entry point. Since perfection is nonexistent, there is no guarantee and no one-size-fits-all design. Nevertheless, doing one's best to ensure safety is never energy squandered.

Nonetheless, what can deter or slow the course of human traffic may provide a few extra minutes to contact the proper authorities. Moreover, visitor passes allow staff to verify who is en route to meet which child and can help determine if permission was granted for such a visit. At this time, I will refrain from discussing parental divorce decrees and visitation, but such circumstances have led to pressing issues.

I favor badge identification and strongly urge the staff to make "badge requests" should a visitor be found without one or if someone is arbitrarily strolling the halls. This strongly enforced protocol builds a more dominant security presence, attesting to the serious intent to prioritize student safety.

In conjunction, I support specific staff vigilance on the interior and exterior of the building throughout the day, checking doors and windows and providing a protective presence by closely surveilling children playing outdoors.

We intend neither to confuse nor create an unsafe learning environment, wishing instead to build a secure juvenile facility. Under no circumstances are we constructing walls or barring windows. The goal is to create a balance to ensure the safety of every student entrusted to the facility.

School resource officers (SRO) are sworn law enforcement officers trained to integrate as a resource for law-related issues allocated to an educational institution. Full-time legal advocates, mentors/counselors, and teachers are a familiar presence at school functions and staff meetings. Independent from the school organization, SROs adhere to their own set of regulations and often provide opportunities for easy communication with the students who realize they are not intent on making arrests and incriminations.

Remember, as parents, you should be well informed regarding the staff and the general safety policies, in collaboration with the philosophy of the school. Attend the meetings and functions, be knowledgeable, and be aware that questions are always encouraged and straightforwardly answered.

CHAPTER FOUR

Law Enforcement Intervention

"Blessed are the peacekeepers, for they shall be called the children of God."

~ Matthew 5:9

What If?

What if posing a question opens the door to panic? How will uncertainty be handled? With common sense or with a sense of risky hesitancy? In truth, it depends on the words preceding the question mark. Let's assume the sequence reads: what if law enforcement does not arrive in time?

FBI statistics document the actions of an armed intruder to run from two to five minutes. Therefore a quick-thinking approach is an essential asset for protection. It is also why I continue to emphasize situational awareness in collaboration with several safety options. The adult who is present at the time of the attack may possibly be the sole decision-maker to ensure survival. the individual who will make the call for a run, hide, or fight plan.

Remember, armed intruders arrive unannounced. They don't extend vocal or written threats, don't pause for dialogue, nor do they spare minutes for deliberations. Most probably the confrontation is a *now or never* scenario with no second chances or do-overs.

Specifying the Nature and Scope of a Threat

In defining a potentially perilous situation, quickly establishing whether the threat is an effective and deliberate intent to cause harm or gravely jeopardize a targeted individual or group is imperative, and also to establish whether the threat is instantaneous and forthcoming versus looming, the most challenging to detect. Nevertheless, in these unsettled times, attention should be given to the severity of intimidation as an oncoming peril to prepare for.

Defining Levels of Emergency Assistance

Before initiating an action, consider the age and vulnerability factors of the children involved. Each age group from three- and four-year-old pre-kindergarteners to college seniors will require a diverse approach. Moreover, it is complicated to predict the reactions of students at various ages once exposed to violence.

Accordingly, consideration should be given to the degree the students involved can understand danger and have the capacity to deal with fears in emergencies. It is also why I am a proponent of practice drills, open dialogues, and familiarity with available resources. Ignorance leaves room for terrorization and panic, the final blow.

We all agree that the best chance for survival is prevention. Be aware—the intruder will not be daunted by the age of the children. Their minds are disturbed; their scale weighing good and evil is off-kilter. Victims of evildoers are often the object of some unknown inner conflict. More often than not the shooting spree is an attempt to even a score. The culprit who angered the intruder will dictate the retaliatory demographic. Bullied at school? A shooter may aim at a class of adolescents. Unrequited love or emotional maltreatment? High school senior girls may be targeted. Unresolved parental issues involving abuse may warrant college-age victims and teachers/ administrative personnel.

Violence is never a resolution for deep-seated conflicts or traumas, but then again, an armed intruder is not reasonable. The more individuals bleeding and lifeless, the more satisfying the task to a person in such a mind frame.

Many times, volatile and rapidly advancing attacks can be interrupted only by law enforcement. However, instructing students and administrative personnel in the trifold options to run, hide, or fight may provide an unexpected course of action that potentially stuns the shooter.

Run

If an armed intruder has gained access to the building but echoes of gun blasts indicate his location is not in the proximity of your location and quick-paced footsteps are out of audio range, it is perhaps wise to head for the nearest exit and immediately evacuate the school. Never hesitate because others refuse to take action. Leave behind all belongings except a cell phone, which should always be in your pocket if possible. Once outside, notify law enforcement. Don't assume the call has already been placed. Never fret over belongings. All possessions are replaceable, not lives. Be fastidious in keeping your hands visible to the police.

Hide

If the blasts of gunfire are eardrum shattering and the aroma of gunpowder is easily discernible, the shooter is nearby. That being the case, evacuation is not an option. As an informed individual, you should have previously scouted the area for safe hiding places. Select a windowless refuge or, if unavailable, darken the room with lowered shades, blinds, and drawn window treatment. Shut doors and, if possible, block with furniture. Make certain you remain in the dark. Remember to set your phone to vibrate mode. All electronic equipment should be silenced. If outdoors, look for safety behind dense concrete or stone walls. Every school should be able to log into Blackboard Connect, a mass emergency notification network through which you can send and receive up-to-the-minute news and alerts via email, texting, and various social media outlets.

Fight

When all else fails and life is threatened, the option for survival at any cost kicks in. If possible, attempt to debilitate, interrupt, or thwart the game plan of the intruder. Age and physical condition permitting, take the aggressor approach, perhaps utilizing whatever heavy object is available, such as a fire extinguisher, chair, or desk. Hurl objects at the shooter. Surprise, stun, and confound. The unexpected paralyzes. Never forget that any loud popping noise can be the result of an armed intruder's actions. Be on the lookout for diverse weapons—knives, bombs, pointed objects. Gunfire is not the only weapon that injures or kills.

Situational Awareness at the Forefront of Survival

Are young children able to make life-altering decisions? When suddenly placed in the crosshairs, there is no room for discussion. A decision has to be taken, and even a child's mind must be prepared and trusted.

How? By emphasizing the importance of situational awareness. Know the exits and windows in a room. Learn to determine which one is best in a given circumstance—to run or to hide in a deep closet or under a nook not immediately visible to a stranger's eye. Youngsters of all ages depend on adult guidance for advice and/or approval of their actions. Be attentive to the proximity of the intruder. Are the echoes of gunshots remote or in the vicinity? Are their voices perhaps muffled behind masks? Or is it a solitary shooter? In some instances, the most detrimental failing is a judgment blunder. Consequently, since all humanity missteps from time to time, it is advisable to err on the side of caution. At most, the result would be a loss of time, not life.

Law enforcement and military personnel undergo rigorous training exercises in preparation for decisions that must be made immediately with little time in the heat of the moment to think—the sole means for survival in certain instances. Familiarity with premade decisions leads to security and confidence in choosing an option while shying away from blindly trailing behind as one of the pack. Identified as herd mentality—the practice of making decisions based on emotion or on what "everyone else" is doing rather than making a rational decision—its adherents tend to be uninformed, often confused individuals who conform by tagging along, believing there is safety in numbers. Unfortunately, serial shooters can down a large crowd within a couple of minutes.

Test cases have demonstrated that in a controlled setting in which a handful of individuals receive pertinent information and detailed instruction whereas all the others are left in the dark, people of all ages from children to adults are more likely to follow the person who seems knowledgeable and eschews self-confidence. The experiment reminds me of playing Simon Says during my elementary and middle school years, an offshoot of Follow the Leader.

The lesson learned is simple. In a pressing situation, the crowd feels safer following as opposed to taking the initiative and responsibility to lead. Nevertheless, Simon's power falls on his followers, those energized to move forward as sheep in a warrior's armor. I typify this *modus operandi* as the easy way out, like a mischievous child facing his father while hiding behind his mother's skirt. What the child fails to realize is that in the end his mother will stay the course as all leaders, and exposed, he will remain vulnerable.

What Is the Most Efficient Defense?

As always, preparing ahead of time is the best defense. Having the right amount of knowledge and information that keeps fear intact and at bay but propels one to attack when necessary. Fear alerts and sets us on our toes, prepared to both defend and/or attack should the situation warrant.

CHAPTER FIVE

Expectations after 911 Call

"People sleep peacefully in their beds at night only because rough men stand ready to do violence on their behalf."

~ George Orwell

What Happens …

"Act as if what you do makes a difference. It does." Wise words from the pen of philosopher and thinker William James. In full agreement, I caution against overloading on fear, a precursor of panic and paralysis usually resulting in deadly aftereffects.

The intruder is inside. Someone has dialed 911. The emergency is reported. Only a foolish individual would believe the threat has been thwarted. So, now what? A hysterical voice shouts for assistance, offering responders a quick, fleeting account of what is happening. Over the radio, cell devices, and police mobile data terminals, the alert is broadcast.

Every school or university should have a safety/survival strategy in place in the event of a serious threat such as an armed intruder.

All individuals outside the facility should be instructed to resist the temptation to respond during such an attack. A predetermined immediate post-911 call strategy should be well organized and easily set into action. Protocol, processes, and procedures should be familiar to all.

I urge facilities, schools, and organizations to have easy access to on-site emergency equipment: backboards, defibrillators, gauze, alcohol for sterilization, and scissors for cutting open clothing to stop bleeding. Decisions need to be made rapidly; lives depend on levelheaded, experienced personnel familiar with situational awareness procedures.

Soon, quickly approaching police officers, deputies, highway patrol, park rangers, and firefighters will arrive—essentially everyone in public safety with a gun or lifesaving skills. Ambulance drivers will remain in proximity on the outskirts until summoned for triage, quick initial treatment, or ER runs in the case of serious casualties and life-threatening injuries.

The most potent enemies in an active shooter event are unnecessary delays and loss of time. Nevertheless, be aware that the first responders who reach the facility will have minimal to no information. Their focus is on evaluating the situation, moving people out of immediate danger, and assessing if the intruder is acting alone. Law enforcement must size up the crisis and enter the premises as soon as possible. Although this might not sound complicated, success is in the timing.

Most street officers don't have the time to canvass institutions. Walk-throughs are not a standard part of their itinerary, although this policy could always change. Instead, the officer focuses on quick action during the crisis, which means hunting the armed intruder(s). Generally speaking, any facility requires the same response. Responders will run to the sound of the gun or follow the trail of trauma until they can locate the shooter. This truth should not instill fear but instead promote an understanding of the possibilities that could occur and to prepare students for the inevitable in an intruder situation. Knowing what to expect in a worst-case scenario will eliminate potentially dangerous panic reactions.

In preparing children and family members, use dialogue that is easy to understand and follow. The topic is serious, and every word should be weighed and presented in a manner to alert, inform, and reassure. The topic is not a light one, but it is necessary. However, once the discussion is open, encourage questions, which can greatly help should the tense moment ever arrive. Well-prepared people can face the situation intelligently and with less room for error.

An officer confronts the situation with a different danger outlook and threat perception. His or her mind is fixated on ending the crisis while incurring minimal casualties. I emphasize the importance of exercising extreme caution. Be on guard—an unidentifiable voice may belong to a crazed intruder, gun in hand bearing a false message of safety to lure his prey into the crosshairs, ripe for firing at a precise target.

Getting into the "Kit"

Optimally, officers should respond in groups of four. But in real life, due to the severity of the event as well as the time and distance it takes for everyone to arrive, first responder officers may have to enter the facility alone. Some may present themselves in uniform, others in bullet-resistant vests, their heads (hopefully) sheltered under Kevlar military ballistic helmets. However, regardless of their outer appearance and garments, each is armed with a handgun. Rifles may be a viable option, but that decision is dependent upon the resources of the responding agency.

The deployment of tactical resources, devices, tasers, pepper spray, and breaching equipment is essential to halt the intruders. When circumstances warrant, officers' behavior patterns may seem all but gentle. Shoving and/or knocking people off their feet is an approach they frequently utilize to remove potential victims from the bull's-eye of a nefarious individual. This is the "kit": "Be safe by choice, not by chance."

Upon arrival, law enforcement personnel's focal point is the entry and location of the shooter. They will also pay equal attention to sounds of gunfire and screaming—a signal of continued threats, fright, or serious injury. Observant of the sounds, officers will head directly in that direction, even bypassing injured, traumatized victims, including children who may desperately reach out for assistance. What seems cold and uncaring is simply a strategy to apprehend the armed intruder before additional fatalities arise. Delay equates to the possibility of additional lives lost and is never a pliable option. The focus is singular: to eliminate the threat and STOP THE BLOODSHED.

Law enforcement has only one alternative—to overlook the injured and frightened, even if they are desperately seeking help. Ghastly visions of anguished, terrorized, blood- and tear-stained faces remain etched in our minds. Training, however, emphasizes the importance of the priorities of life and adopting the rapid deployment mindset, regardless of the role we play or want to play.

I recall not too long ago a time during first-response training when we engaged college drama and theatre students enhanced by moulage, with wounds and injuries artistically decorating their faces and arms. The art of casualty paint when skillfully applied creates a graphic impression. The motive for applying this technique is to prepare individuals and thus prevent shock and psychological trauma when they face a real situation that includes devastating open and bleeding wounds.

Just several days in duration, the training encompasses familiarizing participants with initial response reactions, triage, and medical treatment, as well as the skill set to locate, deactivate, and detonate makeshift explosive devices.

I read not long ago several wise words of unknown authorship that I have espoused as a mantra: “Train as if your life depends on it; one day it might.” As law enforcement officers, we should never lose sight of our purpose to assist at a moment’s notice when unexpected threats loom and to protect and alleviate a victim count before the curtain opens on a devastating scenario. A no-casualty battle triumph is, in my opinion, proof of expert training and the retainment of skills and knowledge. What defeats an armed intruder with malicious intentions? A self-confident armed officer with good intentions and no trigger finger.

The Complex Crime Scene

Meanwhile, in a professional, structured, well-prepared environment, the medical team, in collaboration with public information officers, would assemble an incident command center to guide responders while protecting innocent bystanders, students, teachers, and media personnel flocking to the scene to break the story.

Common sense can be an excellent lifesaver if appropriately utilized. If you believe you are in a safe place, do not wander around the building until reliable news is delivered. Just because the aroma of gunpowder may have waned and echoes of shots dulled does not mean safety is guaranteed. Officers responding to “stop the threat” historically will find the shooter or shooters have taken their own lives or attempted to be put down by responders. It would be a rare situation to "take one alive."

Eventually, the shooter(s) will be identified by witnesses. This signals an all-clear for medical personnel accompanied by law enforcement to take over and start triage or arrange for the evacuation of the seriously injured and those with life-threatening injuries. Photos and videos will be taken to document the crime. Meanwhile, no one is allowed entry into the facility; it is locked down for a nondetermined time frame.

If you are an administrator, I would urge you to plan ahead: Be aware of alternative locations to continue classes/work, and make certain every employee/student file is backed up and stored in a safe place off-site with easy access for a designated individual. Once an intruder has entered, the facility will be a crime scene; therefore, nothing will be allowed to enter or leave the premises. This is my motive for stressing the necessity of information sharing, practice drills, and the hands-on application of a crisis protocol. To do so correctly can save lives.

CHAPTER SIX

Part I
Resolution and Reunification

"When everything seems to be going against you, remember that the airplane takes off against the wind, not with it."
~ Henry Ford

The Protocol

Following an armed intruder situation, recovery and reunification will be required. Procedures should be adopted to assure the creation of a safe area for students and teachers to regroup. Reintegration and reunification are essential requirements that may include diverse phases dependent on professional interventions. Besides focusing on the safety and rescue of individuals and victims, the school must be, in a sense, rescued and able to proceed into full operation without undue delay. Returning to a new normal requires a joint endeavor with law enforcement, a clean-up squad to remove potentially hazardous debris, a repair maintenance team to fix broken windows, counselors, technical personnel to assure databases in the school remain operational, and administrators ready to select a temporary locale to transfer students and teachers so life continues in a familiar pattern.

Though I am a firm believer in the power of knowledge, I have been in law enforcement long enough to realize that much as we try, there is neither a guaranteed way to fully prepare for the worst school nightmare—an armed intruder pointing and shooting a firearm in the face of terrified students—nor is there a means to 100% protect the lives of the innocent and the adults who give their all to educate and assure a healthy safe environment. I continue to preach that we should never halt or camouflage the sharing of information.

Experience teaches that implementing solid strategies can alleviate some of the damage, assure a proper emergency response to triage and treat, and allow students, teachers, and administrators the possibility to suffer less paralyzing trauma when they react with more self-confidence and less margin of error in making life-altering decisions.

Reintegration Process

Catastrophes, whether spur-of-the-moment natural or calculated, can and usually occur as quickly as the blink of an eye. Being caught off guard can result in serious consequences, some of which will have no recourse for recuperation. Random school shootings have engendered public alarm and nurtured the wide-ranging idea that schools should be classified as hazardous for children and teens.

I hope to eradicate this notion through education and by turning different safety options into worthwhile strategies. Among these practices, I favor the creation and implementation of crisis response procedures, with the development of suitable procedures and techniques.

Above all, I recommend shying away from TV/social media exposure during the aftermath, a propagating of information that serves exclusively to traumatize especially young children who regard their classrooms as safe. While the lion's share of media attention usually goes to the event itself, be it a school shooting or another catastrophe, there is generally less attention given to the reunification process, the period in which children are reunited with their parents and guardians. However, as the I Love U Guys Foundation states, "Without a plan to reunite students and parents, we exacerbate the stress and potential trauma of the crisis itself and we jeopardize the responsibility of the school and district in maintaining the chain of custody for every student." Essential for a successful response is meticulous governance with qualified, well-informed individuals in control who at the drop of a pin are ready to spring into action.

The I Love U Guys Foundation is an organization with global outreach that devises and implements crisis response and after-crisis reconsolidation programs. Data reveals that almost thirty thousand education facilities, communities, and organizations have adopted these procedures. The foundation works together with first responders, school personnel, parents, and law enforcement.

A studied and researched reunification procedure, effectively executed, can assist in the dissemination of an emotionally charged situation; guarantee that accurate information is circulated among victims, their families, the public, and the media; and strive to maintain a secure, reassuring environment for children. Devising a well-thought-out plan focused on diminishing the gravity of the situation during reunification is the initial step in the recovery process.

Schools require straightforward, manageable guidelines to streamline more complicated procedures regarding the system of rules during catastrophes, including the classification of a school as a crime scene. In 2013, the US Department of Education and several other government bodies, together with the Department of Justice, the Department of Health and Human Services, the Federal Emergency Management Agency, the FBI, and the Department of Homeland Security, released the Guide for Developing High-Quality School Emergency Operations. The manual was designed to provide a "how-to" model for schools and other institutions, a guidance tool for the formation, expansion, and implementation of dependable plans in the advent of critical incidents at schools.

I Love U Guys Foundation

In addition, the I Love U Guys Foundation, which was created in the wake of the Platte Canyon High School shooting, has developed the Standard Reunification Method. An essential facet of an emergency response is accountable reintegration of students with parents or guardians after a violent event. Safety personnel and crisis responders report to the affected schools and communities offering procedures and tactics to assure a smooth transition into reunification.

Thinking ahead, having a plan, and applying a strategy will also involve collaboration with the interventions of diverse responders partaking in a crisis. Classroom instruction is included, along with workable demos and a step-by-step guide to assure organizations are equipped to reunify students and families.

In creating an overview outlining the crucial components of an effective reunification, I have integrated both the federal government's guide and the I Love U Guys Foundation's protocol, as well as other resources. This is by no means intended to be a comprehensive, step-by-step set of instructions. Rather, it is a summarizing review of the most vital components of a results-oriented reunification process.

My goal is to enlighten school administrators and parents, encourage them to reflect on how they can apply these concepts in their situation, adjust them as needed, and develop their crisis response. It is my hope, of course, that the need will never arise. Wishful thinking? Perhaps, but a positive attitude fueled by a certain dose of foresight planning can promote assurance and resolve a risky situation. Moreover, if necessary, I would like to feel at peace knowing the information I am providing will be a helpful guide for a smoother reunification process that paves the way for a swifter recovery.

Go-kits

The postscript of the 2012 Sandy Hook Elementary School did not read successfully due to a lack of preplanning and a resulting breakdown in the coordination effort, as well as the erroneous quick categorization of the full severity of the situation. As a consequence, the poor chain of command and structural shortcomings led to a system failure in uniting the victims with the appropriate response personnel.

Not being caught by surprise can curtail potential fatalities in a school under siege.

I suggest operating from a stable structure and implementing go-kits in every administrator's office. These are stockpiles of supplies, a vital tool in the event of an emergency such as a school shooting. More detailed than first aid kits, the go-kits should be readily accessible at the reunification site(s), with the district, or at an easily retrievable locale. What exactly is included in a go-kit? Simple:

- Stop the bleed—tourniquet and hemostatic dressing to allow parents to take their child to the ER for triage in the case of a traumatized student terrified to board an ambulance

- Parent check-in banner
- Pop-up tents—quick-open canopy structure suitable for performing lifesaving first aid and emergency refuge
- Folding tables
- Whistles
- Vests
- Dust masks
- Flashlight batteries
- Reunification IDs

Reunification Sites

Never depend on just one site. Circumstances may render a predestined location off-limits or right in the bull's-eye of the shooting. While it may not be possible for all sites to meet the full safety criteria, it's important to understand which would guarantee the most protection and opportunity for assistance distant from the crisis.

If refuge sites are not part of the school district, they should be secured for reunification purposes with a "memorandum of understanding." The purpose of the memorandum of understanding is to set up and put guidelines in place to guarantee a collective understanding among law enforcement, school personnel, and the community of the main objectives of the program with a joint understanding of the different organizational roles in keeping with the safety policy of the school.

Ideally, reunification sites should include the following:

- A spacious parking lot to accommodate parents, children, school administrators/personnel, law enforcement officers, and medical/EMS.

  This arrangement allows for a more expedient system of traffic control on the part of the police and other law enforcement personnel.

- A large building

  The site should provide comfortable, out-of-harm indoor accommodations for the entire student body and staff.

- Private rooms or noncommunal areas

  Private areas for counseling, viewing and identifying deceased victims, and pronouncing death notifications. More often than not, students and teachers may require professional one-on-one post-traumatic stress attention and therapy before being released from the site. The availability of a private area is also a necessary component for law enforcement to conduct questioning and interview students and staff members. Above all, there must be the opportunity to listen to traumatized victims who need the reassurance of a safe reentry to class during this delicate phase of the event. This return is perhaps the most poignant, though crucial part of the recovery process.

- Handicap accessibility

  The building must be in full compliance with the ADA American Disability Act regulations, must be wheelchair accessible, and provide a safety zone for special needs students. Also, it should be in proximity to the school to secure safe, opportune transport without endangering students.

Transportation

Transportation to the reunification site is coordinated through the efforts of law enforcement and incident command. Before boarding a vehicle, everyone will undergo a search and interview process. This gives leeway for other first responders to proceed with their tasks, carefully document and scrutinize the shooting for important clues, and attend to the wounded.

Students and teachers travel by bus to the reunification site. Every individual is accounted for and identified. In some instances, the fire department intervenes in the transport process. It's important to notify parents and staff of the whereabouts in advance. Once known, this location could divert mass numbers of searching individuals who obstruct traffic and create more trauma. Having this location known could divert unnecessary traffic such as parents and news media from the crime scene.

Part II
Training

Active shooter incidents in schools differ from emergencies such as fires. Therefore, I am a proponent of training. How? By following "The Guide for Developing High-Quality School Emergency Operation Plans," which endorses educating students to successfully run through drills as the groundwork for an armed intruder incident on campus.

Of vital importance is meticulous attention to the specific procedures to follow amid a shooting episode. Detailed instructions include emphasizing the situational irrelevance of tending to personal belongings and instead focusing on finding the clearest safe exit areas. Complete and scrupulous training also includes the expectation of the crisis aftermath—reintegration and reunification. These precise steps, if accurately adopted, serve as trauma mitigation while organizing and diffusing emotions at a chaotic reunification site.

Media and Communications

As a consequence of the ubiquity of social media and wireless devices, rumors and incorrect information can spread quickly both within the student body and in the general public. Moreover, the presence of reporters and media teams at the reunification site can proliferate an environment of pandemonium and confusion.

However, imbued with a specific action plan, coordinated with law enforcement and other agencies, enabling public communication can offset the panic, chaos, and disorder that serve to intensify the crisis.

Social media, positively implemented, can deter the diffusion of misleading, inaccurate information. Instead, attention should be focused on the dissemination of factual, truth-based updates. Public information officers should be accessible to govern media interactions, which should not be forced upon students, parents, and teachers until they have received systematic information and appropriate notification of deaths, followed by counseling, should the need or desire arise.

On-Site Services

- Medical resources. Preparations should ensure that students and staff who have been injured or who have special medical needs can receive adequate treatment. All students documented with preexisting conditions should feel confident their needs will be quickly identified and that they will have access to their medications.
- Mental health counseling. A social support system, one of the most efficient therapeutic tools in dealing with trauma and its consequences, should be in place. Therapists, professional clinicians, and rehabilitation advisers trained to offer on-the-spot therapy to handle the death of a classmate

or teacher should also be on hand. Typically, scholastic atavism, sporadic bouts of insomnia with anguishing nightmares, and night terrors can ensue. Every second counts. Professional assistance should be immediate, directed first toward teachers and administrators so they can effectively assist students to appropriately undertake the aftereffects of shock and mixed emotions of anxiety and distress. It is never too soon, once the victims have been laid to rest, to start preparations to ease into a comeback post-trauma strategy. The more quickly a return to the daily norm occurs, the more successful it will be.

- Linguistic rights are sanctioned by every branch of government, and rightly so. Inadequate comprehension only presents obstacles, seeding increased chaos and tragedies. Consequently, attention should be paid to individuals without English fluency. I support readily accessible translation opportunities for non-English-speaking individuals. Schools with non-English-speaking students, parents, and guardians should provide bi- or multilingual communication in both written and verbal form. Contact numbers for counselors who speak the target language(s) should be posted, as well as a hotline phone number for people to receive critical information in their preferred language. Printouts in multiple languages should be distributed.
- Site security. A strong security presence can provide

reassurance in a chaotic situation while preventing further dangers from arising. Preparations should be made to successfully handle parents desperately attempting to find their children. Unthinkingly, many may try to dodge the preestablished procedures. Also to consider may be the influx of journalists and media teams overpowering the site.

The Seven Components of Successful Reunification

Seven basic steps comprise a smooth and successful reunification strategy. Above all are communication and technology. It is important to treat each step as a separate entity, assuring parents and students fully understand the components. If this condition is met, chaos will not disrupt the reunification process.

1. Select and establish a student-parent check-in location at the reunification site.
2. Assign students to an independent area from parents. Each group will be met by "greeters" who will direct them to the appropriate locations and make certain they understand the process.
3. Be sure parents and guardians fill in the reunification cards and the information entered is confirmed. Identification of parents and their children must be a correct. This protocol will avoid parents and guardians from collecting other students during the pandemonium.

- Adherence to the policy will promote "self-sort" during check-in to help streamline the process.
- "Runners" recover students from the student area and transport them to their guardians. Make certain parents are directed properly.
- Controlled, easily visible lines facilitate communication and diminish anxiety.
- Medical and investigative contingencies are in place.

Reunification a Success

Though the seven components look efficient on paper, they will not meet with success without a comprehensive plan inclusive of student relocation, parental identity check, secure entry, and exit points as a starter. Following are other steps of the planning process that can help ensure that reunification goes as smoothly as possible.

Further steps to ensure smooth and safe reunification include:

A plan for transporting students from the school to the reunification location.

- Secured exits and entries at the reunification site to prevent children from leaving before they've been reunited with their parents and to prevent unauthorized persons from gaining access to the building.
- Students divided upon arrival according to grade level.

This protocol is subject to the category of school and the age of the students.

- When and if possible, assign people to the reunification site who are familiar with the students to promote a feeling of reassurance and security.
- Administrative personnel at the relocation site should have full access to district servers and databases for the appropriate and reliable identification of custodial parents and authorized emergency contacts.
- If electronic resources aren't available, the district should provide hard copies at the reunification site for the error-free identification of parents and students.
- School staff, administration, and helpers should be easily identifiable and readily accessible to offer assistance to students and victims. Items such as color-coded vests (included in the go-kits) for quick identification should be available at the site.
- Ensure that procedures are put in place to treat the parents of any of the deceased or injured children with understanding and compassion while honoring their privacy.
- Protocols must be functional and available to inform the school district of any deaths.

Having and following procedures will guarantee a measure

of organization and functionality during a tragic crisis. However, of major importance is the collaboration and unity between school administration, law enforcement, district and school staff, first responders, parents, and students. A successful plan requires all stakeholders to be part of the process.

Made in the USA
Middletown, DE
29 April 2022

64830954R00047